We're Going on a Bear Hunt

LET'S DISCOVER
SEASIDE ANIMALS

This book belongs to:

....

WALKER
ENTERTAINMENT

First published 2020 by Walker Entertainment
an imprint of Walker Books Ltd
87 Vauxhall Walk, London SE11 5HJ

2 4 6 8 10 9 7 5 3 1

© 2016, 2020 Bear Hunt Films Ltd
Text by Walker Books Ltd

Based on the animated film developed and produced by Lupus Films in association with Bear Hunt Films and Walker Productions for Channel 4, Universal Pictures and Herrick Entertainment with the support of Creative Europe – MEDIA programme of the European Union.
Licensed by Walker Productions Ltd.

Created in consultation with Michael Rosen and Helen Oxenbury

Additional wildlife illustrations by Mat Williams/Bear Hunt Films Ltd
Line art by Susanna Chapman

This book has been typeset in Plantin Infant and Dina's Handwriting

Printed in China

All rights reserved. No part of this book may be reproduced, transmitted or stored in an information retrieval system in any form or by any means, graphic, electronic or mechanical, including photocopying, taping and recording, without prior written permission from the publisher.

British Library Cataloguing in Publication Data:
a catalogue record for this book is available from the British Library

ISBN 978-1-4063-9171-8

www.walker.co.uk

All recipes are for informational and/or entertainment purposes only; please check ingredients carefully if you have any allergies and, if in doubt, consult a health professional.

Let's get going!

The seaside is an amazing place to visit. From exploring rock pools to learning about the plants and creatures that live there, this fun guide will help you get going on your seaside animals adventure.

Stay safe

Exploring is fun, but it's important to follow these simple dos and don'ts to stay safe:

- Never go near the sea alone and make sure that any seaside exploring is done with an adult.
- Play safely in the sun and use plenty of sun cream.
- Don't touch animals or plant life and always make sure to put rocks or pebbles back where you found them.

Plan your seaside adventure!

The seaside is such a fun place to explore. Before you go, make sure you've packed all the things you might need for your adventure.

Sticker activity
Colour in your four item stickers and add them to the boxes below.

If it's sunny you might want to pack:
- sun cream
- a sun hat
- sunglasses
- water
- a bathing costume

If it's chilly you might want to take:
- a warm hat
- a scarf
- gloves
- a coat

These items can also be quite useful:
- binoculars
- plasters
- a journal
- a windbreaker

And these beach essentials?
- a spade
- a bucket
- a net
- a towel

A holiday feeling

When you next visit the seaside, take a moment to feel the sand under your feet. Can you hear the sea or birds calling in the sky?

The Big Blue Sea

Sticker activity
Fill in this scene using your seaside stickers.

Fun fact
The sea is salty because salt washes from the ground into river water, which flows into the sea.

See the sea!

As you get closer to the seaside, you might be able to spot the sea on the horizon.

Salty sea

The sea is salty water that covers much of the earth. Nearly all of the water needed for life comes from the sea.

Horizon is where the sky seems to touch the sea.

Seaside scope

Make this colourful telescope to take with you on your seaside adventure. Can you spot anything on the **horizon**?

You'll need:
- A kitchen roll tube
- Two pieces of A4 coloured card
- Sticky tape
- Colour-in seaside stickers

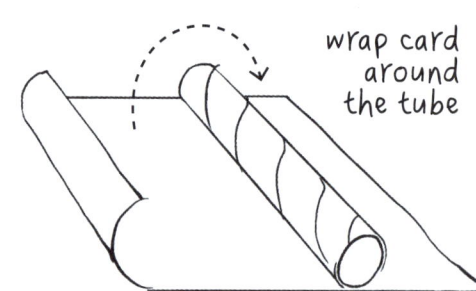

wrap card around the tube

High and low

The sea moves closer to the beach or further away from it depending on the **tide.** The sea washes shells and pebbles onto the shore.

Tide *is the movement of the rise and fall of the sea.*

Sticker activity
Add your shell stickers to the beach.

1. Ask an adult to help you wrap one piece of card over the kitchen roll tube and fasten in place with sticky tape.

2. Take the other piece of card and roll it into a tube shape, making sure that it fits snugly inside the kitchen roll tube. Once you're happy with the size, stick the rolled tube together with tape and carefully push it two thirds of the way into the kitchen roll tube.

3. Decorate the outer tube with your seaside stickers and use your telescope by extending the inner tube.

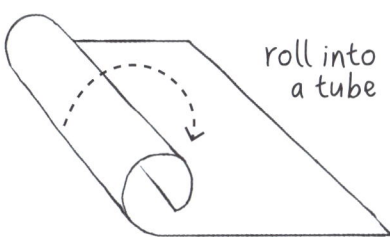
roll into a tube

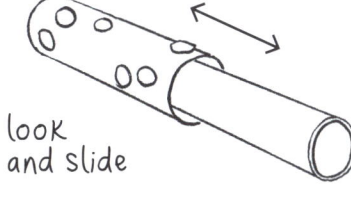
look and slide

The water's edge

The seaside is home to many creatures that live near and under the sea.

Birds can be spotted resting on the surface of the water and crabs can be seen scuttling by the shore.

Herring gulls

These seabirds often perch near the water. They have long slim wings to help them fly.

Thirsty fish

Some fish can only live in sea water. They drink the salty water and pump the salt back into the sea.

Joke corner
What did the beach say when the tide came in?
Long time no sea!

Salty fish

Take the seaside home with you by making these salt dough fish decorations.

You'll need:
- 250g plain flour
- 125g table salt
- 125ml water
- Baking tray lined with baking paper
- Poster paint

1. Ask an adult to turn the oven to its lowest setting.
2. Mix the flour and salt in a large bowl. Add the water and stir until it comes together to form a clean ball.
3. Transfer the dough to a floured work surface and mould into fish shapes.
4. Put your finished items on the lined baking tray and pop in the oven for three hours or until solid.
5. Leave to cool and then paint.

add water
flour + salt

shape the dough

paint your fish

Tough crabs

Shore crabs have a hard shell that can be green, brown or red. They have eight legs, two claws and two small eyes.

Sticker activity
Colour in your crab stickers and add them to the water's edge.

Deep-sea creatures

The sea is very deep and some sea creatures live far below.

Jelly lengths

Jellyfish can be found in deep water. They have soft bodies and long tentacles, which they use to sting their prey.

Activity
Can you think of any other creatures that live in the sea? Can you draw them in a notepad?

Important note
If you see a jellyfish, do not touch it.

Mini jelly puddings

Ask an adult to help you make these mini seaside puddings and add as many details as you can to the beach!

You'll need:
- Digestive biscuits
- A handful of grapes
- Mini cocktail umbrellas
- Blue jelly
- Small glasses

1. Ask an adult to help you make the jelly then pour the jelly mix into the glasses.
2. Drop a couple of halved grapes into each glass. Leave to set in the fridge.
3. Ask an adult to help you bash the biscuits into little crumbs, so it looks like sand, then sprinkle the mix over the jelly.
4. Decorate the beach with a mini cocktail umbrella.

jelly + grapes
bash biscuits

biscuit crumbs

A sandy adventure!

Step onto the beach and see what you can discover!

Smooth sand

Sand is tiny, smooth grains of rock that have broken apart over time. It is perfect for making sandcastles!

Sandcastle flags

Before you go to the seaside, make these mini flags and use them to complete your sandcastle.

You'll need:

Toothpicks
PVA glue
Colour-in seaside stickers
A4 coloured paper
Scissors

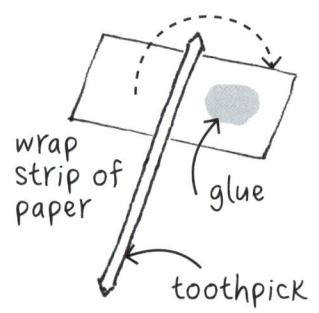

1. Ask an adult to help you cut out a small strip of paper.

2. Wrap the paper around the top of the toothpick and add a blob of PVA glue to the inside of the fold. Hold the paper together in place for a few seconds.

3. Repeat to make as many flags as you'd like and add your seaside stickers to each flag.

A beach game

Use your beetle stickers to play this game of three-in-a-row. The first to get three-in-a-row wins!

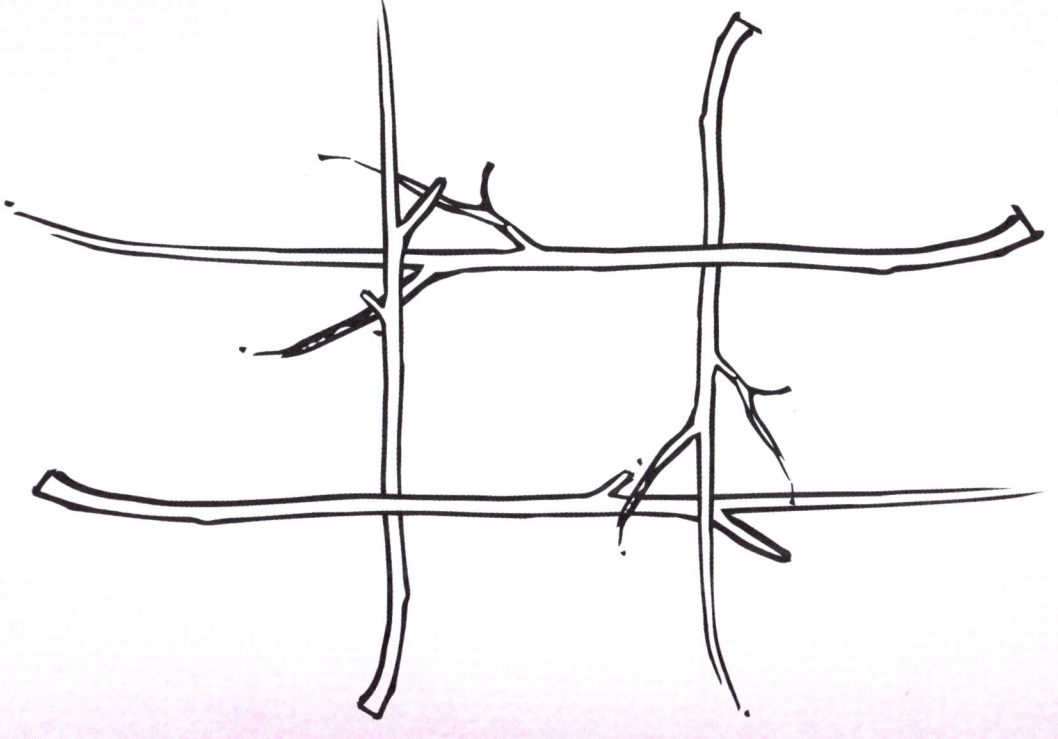

Sand squatters

Lots of tiny creatures like worms, beetles and small crabs make their homes in the sand. Some even burrow under it!

Activity
Dig a hole in the sand – can you see anything wriggling there?

Fun fact
Some seaside creatures, like plankton, are so small that you would need a microscope to see them!

Pool pals

Rock pools are pools of water among large rocks that can be found dotted around a beach.

Take a closer look and see if you can spot any rock-pool residents.

Important note
Always explore rock pools with an adult and only when the tide is out.

Sea snacks

Sea anemones are colourful creatures that cling to the rock, waiting for tiny critters to float by for them to eat.

Keep cool

Make these simple starfish ice cubes and add them to a drink to help you stay cool in the summer!

You'll need:
- A star-shaped ice cube tray
- Blue food colouring
- Summer berries
- See-through glasses
- Lemonade

1. Ask an adult to help you wash the fruit and carefully slice big pieces in half.
2. Add water to the ice cube tray and carefully press a piece of fruit into each shape.
3. Leave the ice to set in a freezer. Ask an adult to help you push the ice out of the star shapes.
4. Pour some lemonade into each glass and stir in a tiny drop of blue food colouring.
5. Drop an ice cube into each glass to make it look like a starfish resting in a rock pool.

fruit

put water and fruit in each star

add star ice cube to lemonade

Super starfish

Starfish are brightly coloured creatures that usually have five arms. They are covered in tiny suckers to help them move along rock pools.

A rocky place

Some beaches have smooth sand, whilst others are made up of pebbles and rocks.

Sticker activity
Add your colour-in seaweed stickers to the rock pool.

Growing green

Seaweed can grow over rocks as well as in rock pools and under the sea. It can be green, brown or even red.

Buckets of fun!

When you're digging in the sand, you may find lots of different rocks and pebbles. Collect them in a bucket and wash them with a little sea water.

Organize your rocks into sizes, shapes and colours. What do the rocks feel like? Are they cold or warm? Smooth or rough?

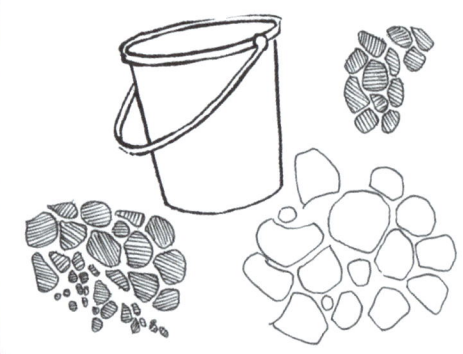

Holiday Homes

Activity
Can you hear any bird calls? Use your binoculars to see if you can spot any seaside birds.

Sticker activity
Fill in this scene using your seaside stickers.

Moving house

Some seaside creatures live in their own homes – otherwise known as their shells!

Home comforts

Lots of shell creatures cling tightly to rocks, waiting for the **tide** to come in to bring their food.

Fun fact
Some shell creatures belong to a group of animals called molluscs.

Activity
There are five shells hidden on this page. Can you spot them all?

Shell sorter

Collect any shells you find in a bucket filled with sea water to wash away the sand. Can you match any to the pictures below? Once you spot one, colour in your shell sticker and add it here.

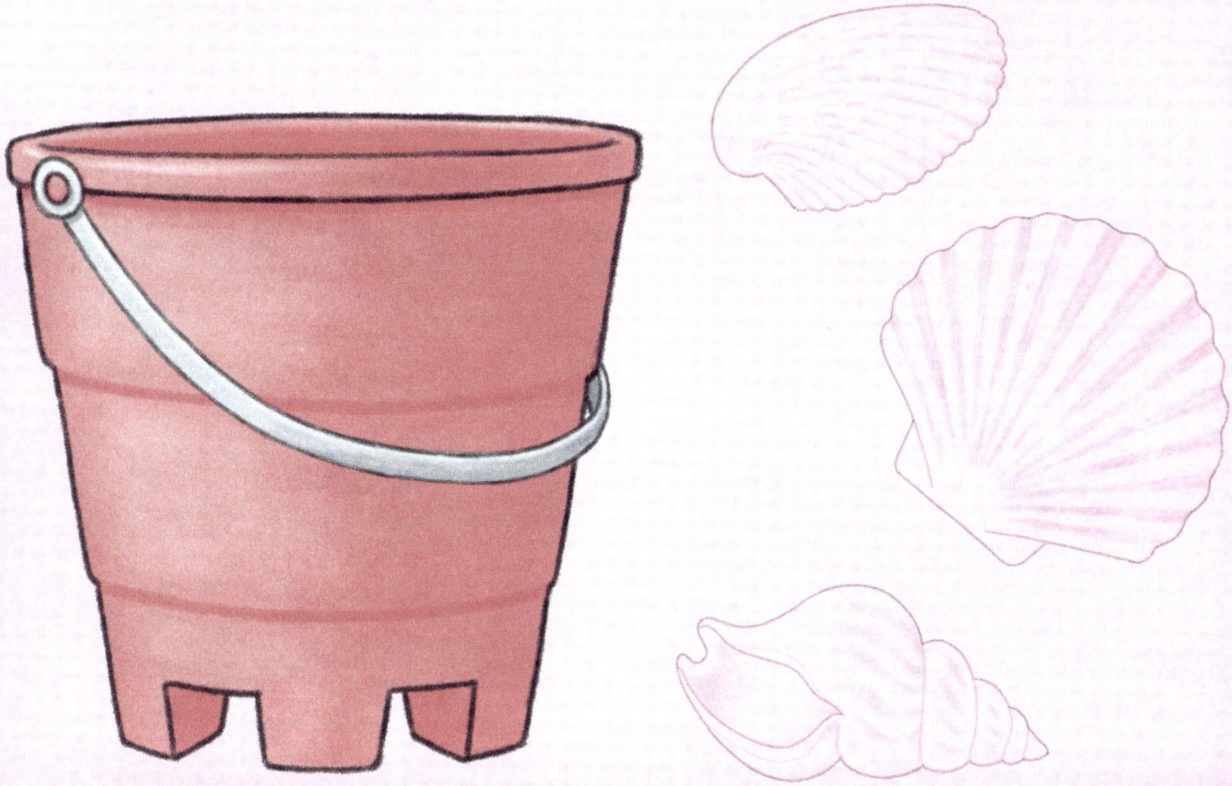

Sea sounds

As the sea moves, the waves can sound quite noisy as they crash together.

If you spot any large shells on the beach, hold one up to your ear and listen closely. Can you hear the sounds of the sea?

High homes

Not all seaside creatures live down on the beach. Some prefer to make their homes in high places.

Sheltered spots

Seaside birds, like kittiwakes, build their nests in sheltered spots in the sides of **cliffs**.

*A **cliff** is a steep rock face.*

Soaring above

Make a simple seaside kite and fly it on the beach when it's windy.

You'll need:
- A toilet roll tube
- PVA glue
- A long length of string
- Tissue paper
- Scissors
- Seaside stickers

1. Ask an adult to help you cover the tube with the tissue paper. Glue it in place and leave it to dry.

2. Decorate the tube with your stickers. Cut long lengths of tissue paper and stick these to the back of the tube.

3. Thread the string through and around the tube and tie a knot in one side. Let the rest of the length of string out and wait for the wind to pick up.

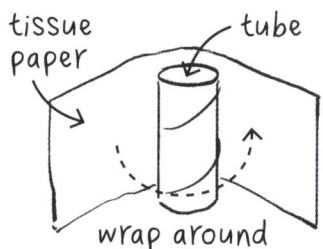
tissue paper — tube — wrap around

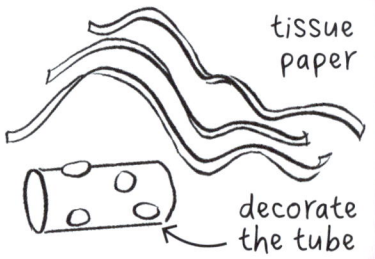
tissue paper — decorate the tube

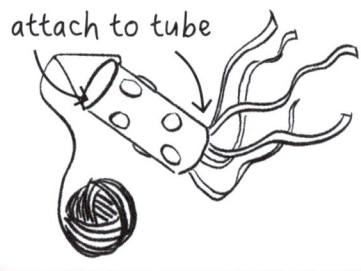
attach to tube

Remote homes

Other seaside birds make their homes much further out in quiet places.

Puffin protection!

Puffins can be spotted living in special seaside **colonies**. They have bright orange legs and multi-coloured beaks.

Colonies are a group of animals living in one place.

Sticker activity

Colour in your puffin stickers and add them to this page.

Curious caves

When the sea washes into the side of a cliff, it eventually makes a hole in the rock face that widens into a cave.

Fish, insects and even bats can make their homes in seaside caves.

A seaside story

When you get back from your adventure, have a think about the things you've seen at the seaside. Can you write a story about your day? Colour in your stickers and add them to this page for inspiration!

Seaside Searches

Joke corner
What is the best day to go to the beach?
Sun-day!

Sticker activity
Fill in this scene using your seaside stickers.

A seaside hunt!

There are lots of signs that animals, plants and people are living near the seaside.

Sandy signs

Even if you can't spot seaside animals, you can sometimes find tracks they have left in the sand.

Activity
Look carefully – can you spot any tracks on the beach?

Handy prints

Use your handprint to make a shape in the sand to play this game of three-in-a-row. If you're on a pebble beach, use sticks and pebbles to play.

1. Draw a grid in the sand with your finger, making sure that there are nine square spaces.

2. Ask an adult to play the game with you. Take it in turns to add a hand-print to each square.

3. The first to get three-in-a-row wins!

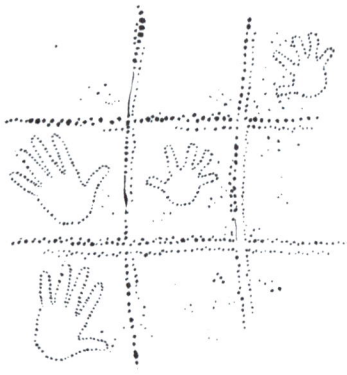

Curlew calls

These large wading birds can be recognized by their call that sounds like "twee-wee" as they fly away.

Make a map

Take a look around the beach. Can you spot any rock pools, a cave, or lots of shells and pebbles?

1. Grab a piece of paper and a pencil and draw a map of the seaside.
2. Use your seaside stickers to add things to the map.
3. Hide a special shell or rock somewhere on the beach and mark this spot on your map.
4. See if your friends and family can find the treasure!

choose your stickers

add them to the map

Sticker activity

Colour in the footprint stickers and stick them along the dotted line to guide the bird back to its nest.

Shoreline foraging

There are so many signs of life dotted around the beach, including driftwood and dried-up plants.

Activity
Have you found any natural objects on the beach? Can you name them?

Joke corner
What did the sea say to the beach?
Nothing, it just waved!

Seaside treasure

See if you can find any of these little treasures on the beach. Once you find one, colour in your stickers and add them here.

Yellow yarrow

Lots of plants, like yarrow, grow near the seaside. This tall, colourful plant can be used by seabirds to build their nests.

Activity

Use your seaside treasures to make a little sailboat and see if you can make it float in a pool of water.

Seaside colours

From the blue sea to the sun setting on the horizon, there are so many colours to see at the sea!

Activity
Take a moment to look at the seaside. How many colours can you spot?

Colour match game

Take a look at the objects below. Can you find things on the beach to match each colour?

Seaside sunset

At sunrise or sunset, the sun is lower on the **horizon** and the sunlight changes to an orange or red colour.

Important note
Make sure that you never look directly at the sun, as it can harm your eyes.

Fun fact
When the sun gets lower in the sky, blue light is scattered, allowing red and yellow light to be seen.

Get packing!

As the day draws to a close, remember to take everything with you! Colour in your stickers to complete the list of things to pack.

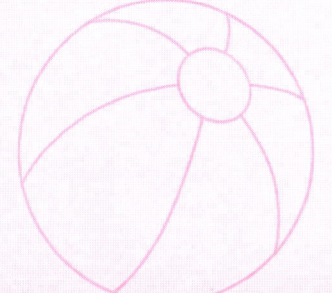  

Fun fact
Crabs can sometimes be spotted scuttling away from the shore as the tide moves. Can you spot any?

Spotter's list
Put a sticker next to any seaside things that you have seen on your search.

The Big Blue Sea

Herring gull

Jellyfish

Crab

Fish

A Beautiful Beach

Seaweed

Spade

Starfish

Bucket

Fun fact
Molluscs can turn their shells different colours by eating colourful foods.

Sandcastle

Ice cream

Puffin

Shells

Kittiwake

Joke corner
How do you cut the sea in half?
With a sea-saw!

Holiday Homes

Seaside Searches

Sun

Driftwood

Curlew

Beach ball

Activity
Take off your shoes! Feel the warm, soft sand or smooth round pebbles under your feet.

Save your seaside

The seaside is a wonderful place to visit and there is lots that you can do to help keep it clean and safe for everyone.

Keep the beach beautiful

Next time you're at the seaside, ask an adult to help you take part in tidying up the beach. Once you've finished your tidy-up, make these seaside badges and hand them out to your friends and family who helped.

You'll need:

- Colour-in stickers
- One sheet of A4 card
- Scissors
- Sticky tape
- Safety pins

1. Colour in your stickers and stick them onto the sheet of card.
2. Ask an adult to help you carefully cut around each sticker so that you are left with discs of card.
3. Turn each disc over and ask an adult to help you fasten a safety pin in place with some sticky tape.

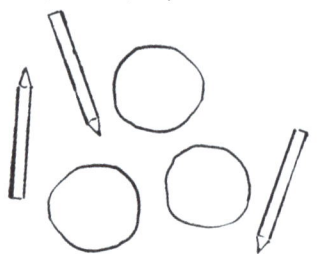
colour in the stickers

add stickers to card

cut out the discs

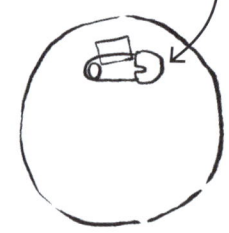
attach a safety pin to the back